Science Matters
CONSTELLATIONS

Frances Purslow

WEIGL PUBLISHERS INC.

Published by Weigl Publishers Inc.
350 5th Avenue, Suite 3304, PMB 6G
New York, NY USA 10118-0069
Website: www.weigl.com

Library of Congress Cataloging-in-Publication Data

Purslow, Frances.
  Constellations / by Frances Purslow.
      p. cm. -- (Science matters)
  Includes bibliographical references and index.
  ISBN 1-59036-410-4 (alk. paper) -- ISBN 1-59036-416-3 (pbk. : alk. paper)
  1. Constellations--Juvenile literature. I. Title. II. Series.
  QB802.P87 2007
  523.8--dc22

                                    2005029917

Printed in China
1 2 3 4 5 6 7 8 9 0 10 09 08 07 06

Editor Frances Purslow
Design and Layout Terry Paulhus

Cover: With a telescope, a person can see stars that are not visible to the naked eye.

Photograph Credits

Akira Fujii, ESA: Page 10R; **University of Heidelberg, ESA:** Page 12T; **NASA,ESA, M. Robberto (Space Telescope Science Institute/ESA) and the Hubble Space Telescope Orion Treasury Project Team plus C.R. O'Dell (Rice University), and NASA:** pages 12 & 13 background; **NASA, ESA, J. Hester and A. Loll (Arizona State University):** page 14L.

All of the Internet URLs given in the book were valid at the time of publication. However, due to the dynamic nature of the Internet, some addresses may have changed, or sites may have ceased to exist since publication. While the author and publisher regret any inconvenience this may cause readers, no responsibility for any such changes can be accepted by either the author or the publisher.

Every reasonable effort has been made to trace ownership and to obtain permission to reprint copyright material. The publishers would be pleased to have any errors or omissions brought to their attention so that they may be corrected in subsequent printings.

Contents

Studying the Stars

In ancient times, stars were easy to see. There were no streetlights, no **smog**, and no tall buildings to block the view. Early stargazers imagined pictures in the sky. They imagined lines drawn from star to star. They could see different shapes. Some saw the shapes of animals. Others saw the shapes of their heroes. Still others saw shapes of tools. People gave each of these shapes a name. They told stories about them. These star groups are called constellations.

■ Patterns of stars are called constellations. *Stella* is the Latin word for "star."

Constellation Facts

Did you know that there are 88 constellations? Here are more interesting facts about constellations.

- Some constellations were first noticed more than 2,000 years ago.

- The shape of each constellation never changes. This is because the stars in the constellation do not change their position in relation to each other.

- Constellations seem to move across the sky during the night, but it is Earth that is moving. As Earth turns, the constellations appear to travel across the sky.

- Today, many stars are still called by the names given to them hundreds of years ago.

- A constellation appears on the flag of Australia. It is the Southern Cross.

The Stars

Long ago, people did not know what stars were. They could see lights twinkling in the sky, but they did not know that they were huge balls of gas in space. They were a mystery. Solving the mystery of the stars has taken hundreds of years. Today, **astronomers** and other scientists still search for information about the stars.

Stars are very far away. Their brightness depends on their size, their temperature, and their distance from Earth. They appear to twinkle or shine because of the light from the stars that is jostled about as it travels through Earth's **atmosphere**. This jostling prevents it from appearing as a steady stream of light.

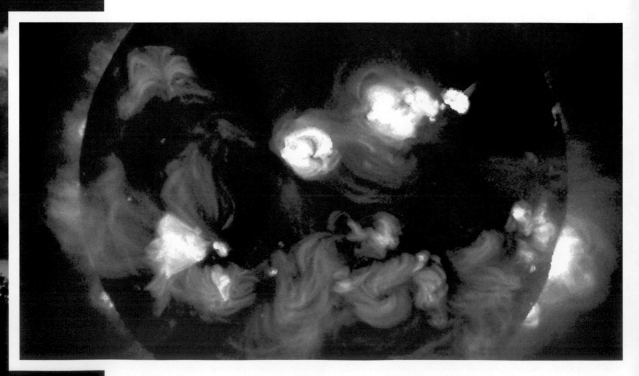

■ The Sun is the closest star to Earth. Like other stars, it is a fiery ball of hot gases.

Constellation Myths

People in different parts of the world grouped the stars in their own way. They told stories about the star groupings. These stories became known as legends or **myths**.

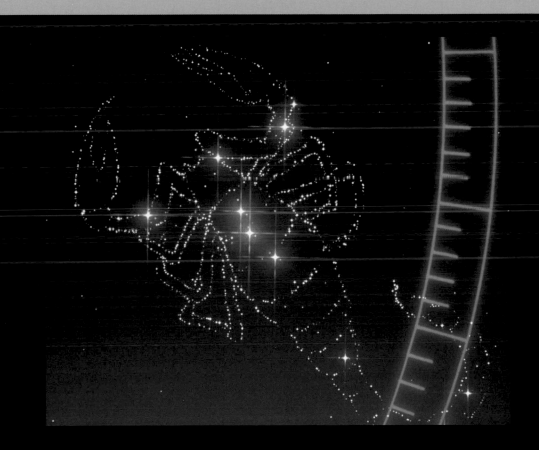

In some myths, Scorpius the scorpion stung Orion the hunter with its tail. Orion fled into the sea to escape. To avoid any more trouble, Scorpius was sent to the other side of the sky. This is why the constellation Orion sets as Scorpius rises in the sky.

Star Map

A star map is like a street map. Instead of the names of streets, it shows the names and placement of the stars in the sky. For hundreds of years, star maps helped farmers know when to plant their crops. Stars also helped explorers and other travelers figure out where they were and where they were going. Slaves in the southern United States in the 1800s followed the Big Dipper to get to the northern states, where they would be free. Today, space researchers, astronauts, and weather forecasters also depend on information from the stars.

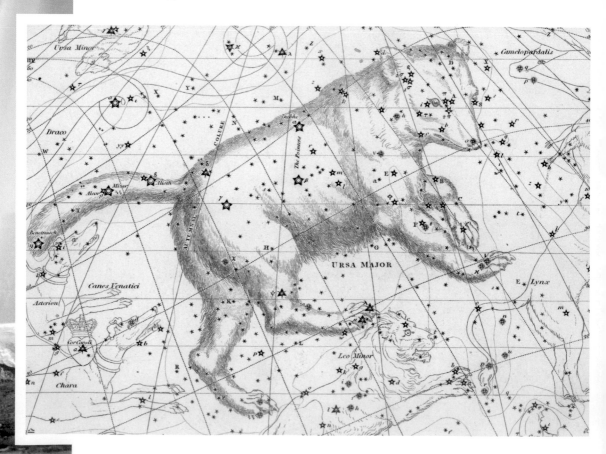

■ Astronomers use constellations as markers to find other stars.

STAR SIGNS

Sirius is the brightest star in the sky. It is found in the constellation Canis Major, one of Orion's hunting dogs.

Ancient Egyptians watched for Sirius to rise before the Sun. From this, they could tell when the Nile River was going to flood. Farmers waited until after the flood to plant their crops.

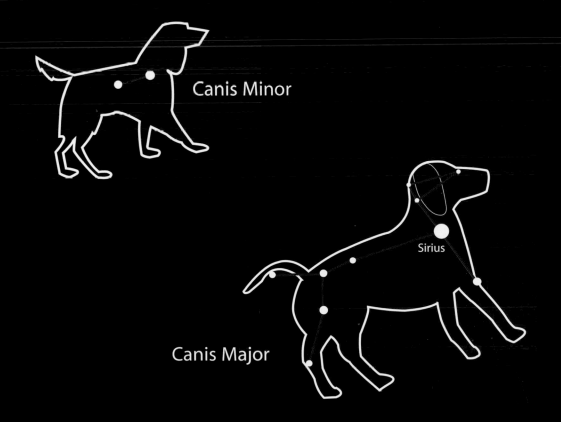

Canis Minor

Sirius

Canis Major

Different Views

A person in the **Northern Hemisphere** does not see all the same stars as someone in the Southern Hemisphere. Some constellations can be seen year round because they are high in the sky. Other constellations are lower in the sky and are seen only during certain seasons. For instance, people in Australia never see the Big Dipper or the Little Dipper. People in North America never see the Southern Cross.

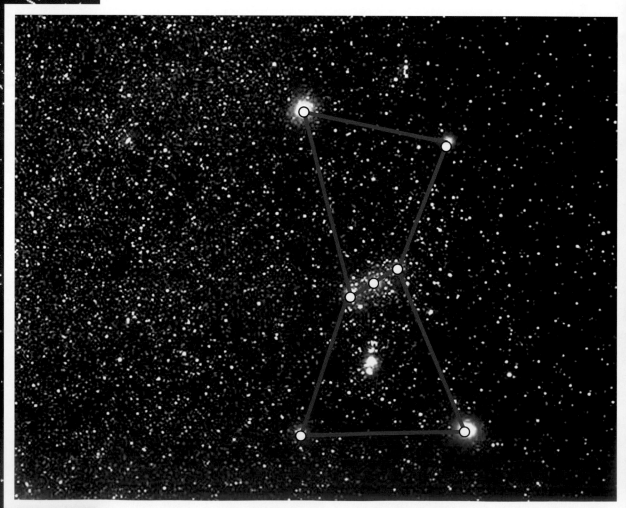

■ Orion is a constellation that is seen by North Americans during the winter and by Australians during the summer.

Finding Polaris

The Big Dipper and the Little Dipper are used to find Polaris, the North Star.

The Big Dipper is made up of seven stars in the shape of a dipper or ladle. It is part of a constellation called Ursa Major. This is Latin for "Big Bear." An imaginary line drawn through the two stars along the side of the Big Dipper points toward Polaris.

The Little Dipper is part of the constellation Ursa Minor, or Little Bear. It is smaller and dimmer than the Big Dipper. Polaris sits at the end of the handle. It is in the same spot every night. Polaris never moves. It is located directly over the North Pole. For centuries, explorers have used the North Star to help them find their way.

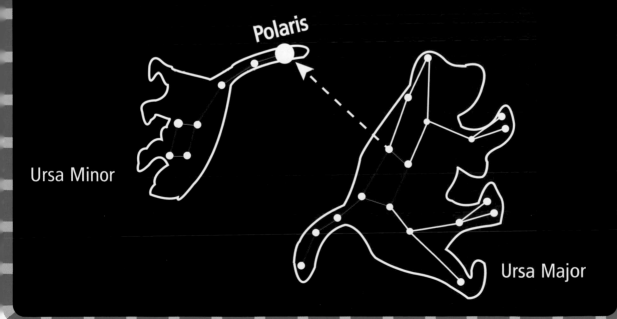

Polaris

Ursa Minor

Ursa Major

Sky Technology

Geographic Information System (GIS)
Special computers called Geographic Information Systems (GIS) gather information about Earth. Scientists use GIS to map **air pollution** in cities and towns. Results are posted on the Internet so people can read about the types and amount of pollution where they live.

Telescopes
Telescopes help us see objects that are far away. Astronomers use them to observe space objects, such as stars, planets, and whole **galaxies**. Telescopes make distant objects appear closer by collecting light. Telescopes can collect more light than the human eye can.

Weather Satellite

Weather satellites are spacecraft that circle Earth. They provide a weather watch on the entire planet. Weather satellites take photographs of Earth's atmosphere. These help meteorologists predict storms and other weather patterns. These satellites also carry special instruments that record information on computers. They monitor events in the atmosphere, such as auroras, dust storms, pollution, and cloud systems.

Radar

Meteorologists gather huge amounts of information in order to predict the weather. **Radar** can tell them what is inside a cloud. This can be rain or hail. Radar can also track a storm that is coming. It helps meteorologists warn people if the storm is dangerous.

Constellations and the Seasons

Constellations are stars that appear to move together. Ancient people noticed that constellations seemed to travel across the sky. They saw that new constellations rose in the east, just like the Sun. However, those constellations moved more slowly than the Sun. It took about 30 days for them to move across the sky. In a year, the Greeks counted 12 different constellations that followed the same path as the Sun. They named these 12 constellations the zodiac. This means "animals in a circle."

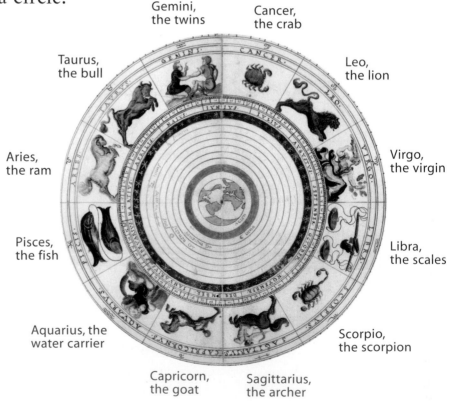

Gemini, the twins

Cancer, the crab

Taurus, the bull

Leo, the lion

Aries, the ram

Virgo, the virgin

Pisces, the fish

Libra, the scales

Aquarius, the water carrier

Scorpio, the scorpion

Capricorn, the goat

Sagittarius, the archer

■ Early calendars were based on 12 zodiac constellations and the 30 days it took each one to move across the sky.

The Zodiac

In the past, some people believed that stars held magical powers. They drew charts to show the position of the 12 constellations of the zodiac and assigned special qualities to these formations.

Today, some people still study the zodiac to show how the position of these constellations control events on Earth. This is called astrology.

Astronomy

Astronomers are scientists who study the skies. They use telescopes to observe stars, galaxies, planets, the Moon, and the Sun. They also collect information using special cameras and computer equipment. Once they gather the information, they **analyze** it. Then they compare their results with current **theories**. If their results do not support the theories, astronomers try to improve the theories.

■ People who work in museums, planetariums, and space agencies rely on astronomers' knowledge of the universe.

Famous Astronomers

Astronomers have made many exciting discoveries.

DATE	DISCOVERY
350 BC	*Aristotle* believed that the stars were in a sphere, like a hollow ball, with Earth in the center. The sphere moved around Earth.
140 AD	*Claudius Ptolemy* collected the world's knowledge about astronomy and created a book. He said that the Sun circled around Earth. He also gave some of the stars their names.
800	Arabs began developing instruments to study the stars. They made the first star maps.
1530	*Nicolaus Copernicus* discovered that the planets circle the Sun.
1609	*Galileo Galilei* built the first telescope to the stars. Using it, he saw moons orbiting Jupiter and the rings of Saturn. He also saw mountains on Earth's Moon.
1660s	*Isaac Newton* designed a new type of telescope using mirrors instead of lenses. He also explained the movement of planets.
1924	*Edwin Hubble* discovered many galaxies beyond the **Milky Way**.
1948	*George Gamow* and *Ralph Alpher* formulated the Big Bang Theory. This theory explains that the universe began as a powerful explosion in space about 13.7 billion years go.

Past and Present

The Hubble Space Telescope was launched into space on April 24, 1990. Today, it still orbits Earth once every 97 minutes. This orbiting telescope takes photographs of the planets and moons in our **solar system**.

Scientists also use the Hubble Space Telescope as a time machine. The Hubble telescope can see galaxies that are 5 to 10 billion **light-years** away. Since light from these galaxies takes billions of years to travel across the **universe**, astronomers are actually viewing events that took place long ago.

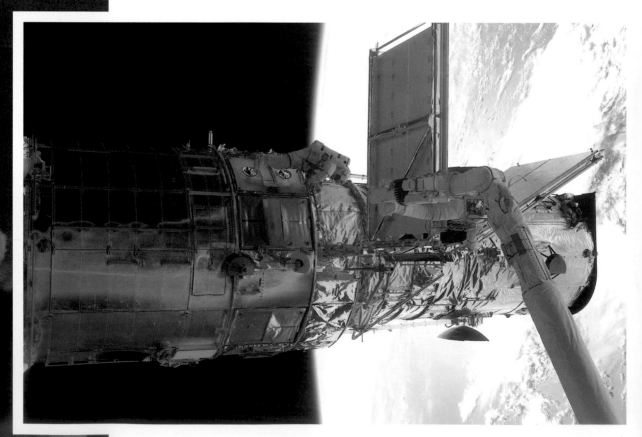

■ Astronaut John M. Grunsfeld upgrades part of the Hubble Space Telescope, which was named for Edwin Hubble.

A Life of Science

Edwin Hubble

In the early 1920s, U.S. astronomer Edwin Hubble proved that the universe is much larger than people thought it was. Prior to his discovery, astronomers believed that the universe was no bigger than the Milky Way. They had seen fuzzy patches when they looked through the Hooker Telescope—the largest telescope in the world at the time. However, they could not explain what the patches were.

In 1923 and 1924, Hubble was able to pick out some stars in the mysterious fuzzy patches. He figured out that those stars were about 1 million light-years away. He went on to discover many new galaxies.

Surfing the Stars

How can I find more information about constellations?

- Libraries have many interesting books about constellations.
- Science centers and planetariums can help you learn more about constellations.
- The Internet offers some great websites dedicated to constellations.

Where can I find a good website about constellations?

Encarta Homepage www.encarta.com

- Type "constellations" into the search field.

How can I find out more about constellations?

NASA Kids http://kids.msfc.nasa.gov

- To see some constellations outlined in the sky, first type "constellations" into the search field. Then click on Chapter 9 Learning About Constellations. Click on the arrow in the illustration.

Science in Action

Make Your Own Constellation

Use your imagination to create your own constellation design.

You will need:

- a pencil
- a shoebox
- scissors
- tape
- a flashlight

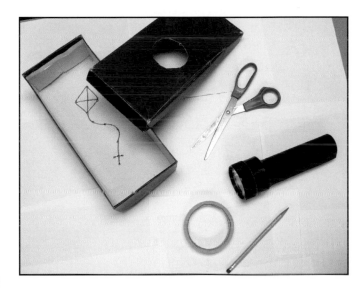

In the bottom of a shoebox, draw a shape that you would like to make into a constellation. Use a sharp pencil to poke small holes along the outline of your drawing.

Ask an adult to help you cut out a circle in the lid of the shoebox to fit the flashlight. Then tape the lid onto the box.

Place the flashlight into the hole and turn it on. Shut off the lights in the room.

Look at the shape of light dots on the wall. Can you see the shape of your constellation? Make up a name for the constellation you have made.

What Have You Learned?

1 How many constellations have been named?

2 How many constellations are in the zodiac?

3 When was the Hubble Space Telescope launched into space?

4 In which constellation is Sirius found?

5 Which constellation is the Big Dipper part of?

6 How does the Big Dipper help a person locate Polaris?

7 Why do we see some constellations at one time of the year, and not another?

8 Who built the first telescope for viewing stars?

9 What does *stella* mean in Latin?

10 What does an astronomer do?

Words to Know

air pollution: harmful materials, such as chemicals and gas, that make air dirty

analyze: to study carefully

astronomers: scientists who study the universe

atmosphere: the layer of gases that surrounds a planet

galaxies: large groups of stars

light-year: the distance light travels in one year

Milky Way: the galaxy that Earth is part of

myths: stories passed down for generations, often about gods

Northern Hemisphere: the half of Earth north of the equator

radar: a system that uses radio waves to locate objects

smog: smoke mixed with fog

solar system: the Sun and everything that orbits it

theories: ideas of how or why things work

universe: all of the stars, planets, moons, comets, etc., that exist

Index